THE ELECTRON-GHOST CASINO

THE ELECTRON-GHOST CASINO

RANDOLPH HEALY

MIAMI UNIVERSITY PRESS

Library of Congress Cataloging-in-Publication Data

Names: Healy, Randolph, 1956– author.

Title: The electron-ghost casino / Randolph Healy.

Description: Oxford, Ohio : Miami University Press, 2024.

Identifiers: LCCN 2023027127 | ISBN 9781881163732 (paperback)

Subjects: LCGFT: Poetry.

Classification: LCC PR6108.E15 E44 2023 | DDC 821/.914—dc23/eng/20230616

LC record available at https://lccn.loc.gov/2023027127

Designed by Crisis

Artwork on page 89 by Genevieve Healy

Printed in Michigan on acid-free, recycled paper

Miami University Press

356 Bachelor Hall

Oxford, Ohio 45056

CONTENTS

PREFACE: WHAT ARE YOU AT?

Well, you tell me. I'd like to think that anything *can* be a poem, unless it doesn't want to be. Or that a poem should be at least as interesting as a good conversation. I've been informed that what I write isn't poetry. Suit yourself. I've also read several statements which have been incorrectly attributed to me. For example, that I am part of a neo-avant-garde (No, but thanks anyway), or that I don't consider myself to be an Irish poet (Who *wouldn't* want to be an Irish poet?), or that, poor, ticketless item in poetry's lost-and-found, I fruitlessly seek recognition (Nonsense. My wife knows who I am, and so do I quite often).

The title, "The Electron-Ghost Casino" refers to Rutherford's gold foil experiment of 1911 showing that atoms are mainly empty space. If you took away the empty space, the entire population of the earth would fit in a teaspoon. Of course, you wouldn't be able to lift it. (I should credit a student of mine, Logan Duff, who asked "Are we all then just electron ghosts?" when we were covering the topic.) The appearance of solidity when Samuel Johnson kicked a stone in a bid to refute immaterialism was due to his electron fields repelling those of the stone. The casino bit is a gloss on evolutionary processes, where most trials end in failure.

Some declare that poetry must be simple. That can, paradoxically, lead to texts aimed at imagined beings who never read, or to works of effortless brilliance. Others swear that poetry should be rebarbative, politicized, subversive. Here, the gamut runs from the dull but innovative to the exhilaratingly engaged. If these are endpoints, there might well be a vast continuum of possibilities between them. There are over six billion ways to rearrange the words in this sentence. Anyway, my thanks to the first hominids who proto-linguistically gargled to achieve ecstasy or to instruct and delight their peers. We have come so far, from caves to the precipice. *Vivi felice.*

RANDOLPH HEALY

THE ELECTRON-GHOST CASINO

PEGASUS X 2 AT POWERSCOURT

Students bob up and down, boisterous in thickets
of language they think foreign to us.
A man and a woman, leaders, puzzle over
*Berlin Hugo Hagen fecit *869*
and mile upon mile of emulsion is permanently altered
by light bouncing off dazzling smiles.
"Fabless," he says, "fabless," while she, behind guide book,
motions left, so left they go, leaving the zinc chargers,
rain-stained wings straining against stillness,
not-veins not throbbing on their thighs and bellies.
Images of perfection do not trouble
the breezes that riffle life's lexicon.
Generations rattle past like notes
in a robin's call (run your fingernail
along a comb) while the wind still gusts
species streaming like bubbles from its touch
each rainbowed shape (including ours) a prelude to another.

THE TENSILE STRENGTH OF LAST WINTER'S ICICLES

I.M. ED DORN

a cool head takes time
tossed the dude

      but a cool time gives you head
parried an equine if unexploded blonde

O Miss Polly
give me an infinity you can count on

or at least a decent tattoo

in return take my retinas
(though I'll keep the negatives)

            careful!
            they ain't no
            things!

then there was the sound
of something natural

who's there?
    "Jus' wee me,"
said the cheery sower

trust a German to write
Being on Time

this is for your sadly missing heart

SOLATION

An aisle in a basilica
where camellias bless absence
and coal near a chasuble
sears bluish billows.

Wisps in the cupola
eclipse the will
inordinate sorrow
its own exile.

Shall we save a tomorrow
glittering with bodies
harvested from distant breakers
when ill shall waft shafts of ash
and filial light list and fail?

OUTTAKES

One must be limp a long time
not to mob the tanned and maimed.

Surround the de-nippled torso
arms out as if no one was more willing
unworked hands judged beautiful
whorls spiralling beyond naming

then rubberneck a neck-wrung ex-sixteen-year-old,
Munch-gape made of meconium
as if she had shit herself
or someone had sat on her while still warm.

Don't miss the Iron-age upper crust
who'd mussed a coif with French pine resin
to be attacked by nits and axe
everything from the diaphragm down still missing.

A sign says they have no identity
peat having leached their DNA.
Someone wonders how much they'd go for.

A woman who ferried the torso by car,
obsesses about arms
reaching to get her.

Finish with a pair of ex male bonders
de-cocked, de-noodled
their own tripes as neckties
to exit as altered
as one who's looked over
a celebrity pickpocket's
wallet collection.

TWELVE RUSSIAN DOLLS

FOR MAIRÉAD BYRNE

a fatling entitled I
said fling tinted die
ideas if tide denies
disease if id defines

see physical I germinate
sense apishly manic regiment
screens play iceman eating
recession a masculine agent

rot this splintered tongue
trio hints epistle gone
erotic things spite ego
creation lightens its o

VENUS AND MARS BARS

FOR RACHEL LODEN

"Twenty little buddies in a box,"
she said lighting with a swagger.
"Get a load of this gargoyle
(nice wheels for a beginner)."

So I pulled over and savoured
her coffin-nail halo
as she arched like a caterpillar
testing mimosa.

"Give me a dog any day,"
said a passing rocket scientist
"Train it to pull a lever,
it pulls the lever.

"Those bloody chimpanzees—
couldn't keep their paws
off a thing—it's a miracle
we got any of them up."

The gang on her ankle chain clattered
as she disappeared in a taxi.

That old flirt Feynman said that if
by some cataclysm
only one sentence could survive
he'd choose, "All things are made of atoms."
So I bought myself a fun-size
and fizzled to its $C_6H_{12}O_6$.

UMBILICUS DISCONNECT

(IN RESPONSE TO A POEM BY CRIS CHEEK)

Chew the scud
Urged a pasteurised sop to a puzzle
Each precious stub
Unveiling with gusto an infra rouge infection
Unnoticed by muscle
Scrabbling for quarry
Who would not tussle when so utterly funded
Mine mine assails the ingrate
Equating the adult with unusual outrage
(A form of abandon unsuited to uniform)
The stars and the bars require super vision
When anticlockwise is hardly an option
And those who interrogate
Get voted off the nation

CUP AND LIP

When news of Christ first came to Greece
quite a few misheard his name as *Chrestos,*
the useful one, and thinking him
some kind of zealous plumber
gave the meetings a miss.

Invincible Ignorance, I ask was this
the lightest ever road to paradise?

SPHINX OF BLACK QUARTZ JUDGE MY VOW

"You catch more flies with silk,"
said the actress to the scalpel
cutting loose a toast rack
chair-strapped for being sexual.

Barbequeen never feels blue.
"I built myself," she chirps
"from a tiny little starter kit.
Well, Mom helped a bit."

Freedude improvises an RPG
by turning the taps on full blast.
"My queen blows hot and cold," he chants
"Her golden apples blighted at last."

Futures burn rubber
as histories nurse their hutch burn.
The visitors' book lists ghost after ghost
every letter of the alphabet rubbed out.

ASHEN RIOT

resonant tattoo
as seas shatter streets shanties
so nations thirst in tents

onanist saints
toast atheists' intestines
to atone

ores tarnish
in a senate that is theatre
as inheritors

restrain or shoot resisters
see no as treason
anaesthetise treaties

terror is here
stress haste
and soothe assets

roses thorns
o reinstate an interest in stretto
raise a retsina to other striations

PAPER CUTS

FOR KEITH TUMA

Ok so it hurts, but what doesn't?
Now the ECB's run out of ink
we have to imagine what each note's worth.
(Makes tipping more stressful.
What's fifteen percent of what?)
Look, another striker in a cast.
Then the power failed and Tinkerbell was substituted
by a lighted match taped to an ice cube.
I wonder how long the match lasted.
Oh well, magic isn't forever.
As part of the team-building thing
we all had to get into this enormous fucking canoe.
It was so cumbersome
we couldn't get it anywhere.
All the same, I'm glad we've a smart economy.
Though part of me wishes I'd invested in brass.
Maybe gold, silver, iron or lead.
I'd certainly think about joining a party,
or at least a huge human pyramid.

(Though if you somehow make it to the top
it's still not going to go
anywhere it doesn't want to.)
Neck, heart, tongue, soul, arse.
Practically a robot.
He said it wasn't worth his while working.
She said maybe it'd be worth someone else's.
That killed the conversation.
He stubbed his fag and started playing
the electric organ till she pulled the plug.
Tickling the plastic isn't the same.
I read it online.
But what I'd really like is an ebook
with predictive text
that could improvise according to your mood.
Or, better still, according to the needs of society as a whole.
I'm definitely joining a party.
Hello. Yes. Not yet.

THE ROAD TO GOD KNOWS WHERE

One warm evening
when the air reached into
our clothes as if nature wished us well
we wandered over Breton dunes to find quiet
shredded by cacophones
accompanying a troupe lurching about
in garish eiderdowns.

Take a cheese grater to your cortex
or eavesdrop on the god of thunder kazooing Dumbo
followed by the inter-county T-Rex throat-clearing competition.

Keats says in a letter to John Taylor
that *if poetry comes not as naturally*
as the leaves to a tree
it had better not come at all.
A sudden drop in temperature triggers leaf-fall.
No, it's the shortening day.
But if we from natural processes result
is it possible for us to be artificial?
Stars are much easier to understand than trees.

Once Frank could just glance at a map
and head straight there
in a wrong-hand drive anywhere.
Then in Portugal he had to ask Cora for help,
and next on the way to Dromoland
he got lost entirely.
The consultant said that it is very rare
for the vascular forest to lay such waste.

There is nothing to say
and nowhere to walk
the invisible strings cut
and music and thought are lost.

CARBON FOOTPRINTS

I needed a bed so
my mother bought a book on D.I.Y.
and gave it to my Dad
his first project my repose.

How I loved the smell of fudge
drifting from his Dutch tobacco
and the touch of his sea-foam
-frozen Meerschaum.

Each night when I made my bed
I thought of him
sometime around 3 a.m.
when it collapsed again.

Eventually, I slept at one end
of a mattress on the floor,
my brother, Ed, at the other.
The flames didn't wake us

unlike Dad stamping them out
with his bare feet before
turfing our bedding out the window.

Figuring that Ed at seven
had had a go at smoking
we went back to sleep
to wake again

the night sky lit up
by our blazing coal shed.

Mary handed me Dad's casket
which I put in the ground

agonizing over whether
I'd gotten it the right way round
forgetting that ashes don't have a bearing.

FRANK

"You have to be very tough in London,"
he said, and thought of home, but didn't return
since his brother had sold his house on him.
So he went further south and got a job
fitting engines in a shipyard. But then
he lost it. A smash-up. Mistake really.
Someone told him they were looking for blokes
to go on a seal cull, and he went.
Hard life, but the people were okay.
Though best to sleep with a knife under the pillow.
The seals mostly slipped past him and died
in the water. Skins ruined. Still he got enough.
Off the coast, all finished, the boat sank.
So they waited in port for the insurance.
But the captain ran off with it and
someone stole his passport. Then he lost his head
and skipped the border to America.
Deported back to Ireland. Slept rough
for a while then a priest found him and got
him a room in Brendan's. Never was married.

THE GIFTS OF THE SPIRIT IN THEIR ORIGINAL PACKING

"It's great to be alive,"
said the long-lived man
sticking 85 photos of himself
to his window, facing out.

His spitted son lurked by the lavender
still smarting from a run-in
with a raven-haired artist—
fatherly, gifted, a looker.

Earlier, a clatter of stilettoes
chatted as he whittled,
he lit by their giggles,
they frozen as he pranged
a guinea pig one was toting.

He showed his face then headed off,
the wind singing,
"Count the rings
when you're cut in half."

MINOAN MINIATURES

FOR GAVIN SELERIE

LENT

Eaten transubstance is soon forgotten.

TRÈS LUISANT

This clear-sky New Year's sun
this disc incised in scattered blue
from which radiant barbs enter
tender skull-mounted orbs
which next to nil assimilate but gorge
on torrents of retinted reticles
indeterminate clarity
sparking lightning traceries.

QUESTIONNEZ

Milhaud asked Satie, the latter getting drenched
on his way to Montmartre,

why he wasn't using the tightly rolled umbrella clutched
under his arm. "Are you mad?!" Satie answered,
"This is far too valuable to get wet!"
After his death friends found six identical velvet suits
and dozens of umbrellas in his flat.

DU BOUT DE LA PENSEÉ

Without lacunae a ladder is just a plank.

POSTULEZ EN VOUS-MÊME

Manzana in corpore sano.
An apple a day keeps the doctor away.

PAS À PAS

Static erased your nascent song.
O opaline world abrupted, for what?
No reason.
Ghosts orbit full tilt in vacuo.

Under, a furnace is frozen.
Nadir, I implore you
Send grief, hurt, rage.

Unlevel this spirit
Necrotically making do.
Grace it with agonies to feed.

SUR LA LANGUE

Sinuous, sliding source
of silly and celestial,
darting, lashing, tale-teller,
print unique as those digital,
caruncles secrete,
silver devils,
Jas. 3: unruly evil, sink of iniquity,
slitherer, roaming such little labyrinth,
seeker of fragments,
testing each crevice,
ticklish sensories throng thy terminus
on which little skillet salt doth sizzle,
saluti.

PEEL ME A FRUITBAT

You can't call it help
if it always ends in a barney

smell of hot fat
lenses frosted by sea spray

met us a
woman with serpentine plaits
stone beautiful
until the

thread snapped

better buy a lot of ballots

or cast a god
one-eyed
into a forest
where a tree groans

and the last crate of light
heads back to base
with not a tosser saved

faster faster

go the lost and clocked
feeling
what scars feel
long after departure.

CRASH

My computer died during the week and all the work I've written in the last year and a half has gone to another place, which I'm not entirely unhappy about. I'll try to reconstruct some of it.

One was about monkeys, lots and lots of monkeys. They were using an older technology and they had a lot of time on their hands, producing texts that contained every detail of every biography real or imaginary every thought or dream or urge and so forth.

There is a site where a guy has set up electronic monkeys and you can join in and your computer can be part of the zoo. It's happening all over the world and they're generating texts in this manner. So far they've got a three word sentence.

Some believe that it's not possible in the time available for randomness to generate a world as complex and interesting as the one we inhabit. I have mixed feelings about this. Perhaps the process started earlier than time. Though Saint Augustine said it doesn't make sense to ask what happened before time began because before is a temporal word and you beg the question.

I've a feeling that this anxiety about determinism and randomness is not going to go away. I don't know what randomness is anyway. Many of the definitions are concerned with how much we can predict or understand of a given system, but if it had any constructive role in the universe it must have been there before human consciousness.

The difficulty I have in reconstructing anything arises from the last time my mother polished the floor. I was about four at the time, there must be a genetic factor in this, because I didn't like wearing shoes and none of my children do. I ran across the floor, and slid extremely fast into the door frame, fracturing my skull. I woke up in my mother's arms and she was saying "It's all right," over and over again. But it wasn't, and it felt very unpleasant, such intimacy.

(phone rings: the ringtone is from *Gran Vals* by Tarrega)
(it rings again)
It's my wife. Hello.
"How're you doing?"
I dunno. I'll ask. How'm I doing?
(various answers)
Not too bad. Not too bad.
"You're not doing your flippin' reading are you?"

I am.

"Will I leave you then?"

OK, sure.

"One thing, do you know where the sheet protectors are?"

There were some in the top shelf above the hot press.

"Ok, bye then."

Bye.

I grew up in Finglas where you didn't actually go to the doctor so I'm not sure what the consequences of that were but I do know it's led to depression, in a literal sense. You can put your fingertips into various parts of the landscape of the back of my skull. It's also led to complete blanknesses. One of the bits I really like in Frankenstein is where he goes up to the North Pole and there's absolutely nothing. I can have whole days like that. It's very relaxing.

There was something about viruses. If you look at the human genome it's absolutely full of viral DNA as if we were some kind of dumping ground for them. Anyway, even if you don't want to go one side or the other, determinism or randomness, things aren't going to stay the same. We are human beings at the moment, but it's not gonna last. Then there was a punch line that was very sharp, but I've forgotten it.

Another one was about a hermit and he lived in a cave underwater. Spiders used to come down with bubbles on the hairs of their legs and fill up the cave with air for him. He was a very holy man and this affected the vegetation in the cave. Some of the algae began to manufacture luciferin, which gives off light, and sacred texts began to illuminate the walls of the cave. Hermits these days are very solitary kinds of people. But in the middle ages it wasn't like that. One, Peter the hermit, had a woman walled into one of the sides of his cave. She seemed to be quite happy with the arrangement.

There were others.

THE OM IN REMOTE
THE NOT IN CONTROL

Polar bears in synchronised drowning
lights going out all over a galaxy
we can't choose then out of the blue
drag up Carl and Susie.

Countdown began when
he faltered and left his knight *en pris*
and she led him, broken,
from the tourney.

"Go out with a worm," he said,
"they're full of hearts"
then spent the night in a car park—
no fuel, no money, the power cut.

Eyebrows raised when days after
she dumped him he was copped
at the copier, all of a frazzle,
faking a birth cert for an offshore chattel.

HOLD YOUR BREATH FOR GLOBAL WARMING

FOR DECLAN O'NEILL

Wriggle like a snake
waddle like a duck
that's what you do
when you do the hockey stick.

D-I-S-C-O_2

D-I-S-C-O_2

The angelic legions weary
of sublunar mire and would return,
their tessitura fractured.
Flung from the celestial sphere
the gloried clouds a buckler
against their redemption,
though they broil the earth
they are cast out.

When the sun hits your eye
like the square root of pi
that's statistics.

When you call a result
based on seconds of a game
that's prophetic.

Unless they're the dying seconds.

DUET

FOR CANDICE WARD

She made seed hoppers
in her workshop in the attic
and every new moon
hired a cherry picker
to hang her latest from the apex
of a sexually mature alder
returning to write a love letter
which she tore in two
sending one half to the other side
of the earth, one week later
putting a match to hers so
that the other would also burn
though she, like the alders, disliking
summer heat, went north each June
finding it easier
as she walked in the cool of the day
male catkins relaying their restless drifts.

WILLY

My father was a writer too
working right on the edge
maintaining an amorous correspondence
with a large number of different women simultaneously
until having repeatedly placed one person's letter
in another person's envelope he shorted the entire network
down to one final pen-friend in Scotland.
Leaving nothing to chance,
he got on the boat and asked her to marry him.
"I suppose so," she replied,
and they did, he returning to Ireland, she remaining.
My sister was born, then I,
so she moved to Dublin, her heart not in it.
He wrote ballads too,
one in the person of a mother grieving for her son
cut short by means of capital punishment
the refrain being
They hung my Willy on a tree.
I aspire
to leave as worthy a legacy
to my children.

PAPER WINGS

Teddy is standing
keeping as still as he can
with one arm out
on which a moth has landed
which he says he's going to hand-feed
to one of the bats
swooping round his tent.
Beatrice is not impressed.

Have we all grown up then
surrounded by accounts
of imminent annihilation?
Marie-Louise worried about
the plants and the animals—
that nothing would be left
save silicon creatures by deep-sea vents
and MRSA in hospital rubble.

Here are some lines I crossed out.
~~If whales are so clever—~~
~~why didn't *they* discover fire?~~

Maybe they're too clever.

And

~~Thus god created Lucifer~~

~~whom he then expelled~~

~~cleaving a hole in his very heart~~

~~into which I must presently fall.~~

I don't understand.

Not to mention

~~"Let there be peace," whisper scouring streams~~

~~as hellfires descend from unstable drones.~~

But to wrap up I could say

that something is heading in our direction.

But not with a straight face.

GRAND GESTURES OF THE IRRELEVANT HAND

It's hard for questions
when pro-forma answers swarm
a flash of light a crash
a scantily clad assistant
it doesn't take magic
to make inquiry impotent
but truly it would be easier
to send a shoal of sharks
to extra-solar inquisitors
than to jail percenters in this jurisdiction.
O children, and your children's children,
indentured spare parts in these fixers' boneyards,
here are your millstones, the water's fine.

MORNING CAME DAILY EXCEPT WHEN THE NIGHT WAS LONG

Pick the wrong piece of pottery
and it's over pal.

One tribe in the South have a dance
that encodes a recipe
for a particularly delicious soup.
It's not something they do at weddings
but it's a good dance for a cold winter's night.

Spiderman, wearing his best costume,
went over to the Fantastic Four
saying he'd like to join.
"There'd still be as much alliteration," he explained.
"We're good with four," they said.
He lost a lot of weight after that.
In the end, his dead uncle appeared and said,
"Look, Pete, do it because it's the right thing to do.
Forget about that floozy and the bad reviews.
And, for Christ's sake, stop whining."

How we loved our old volcano.
Even though it was extinct
we still went to see it.
Respect.

Then a risible tide shaved all goats.
But we didn't even have a snorkel.

At the launch of the world's most inflated democracy, the Placeholder's new scrotum experienced mechanical failure. Cameras were deep-fried and the building wrapped in alufoil. Officials applied an emergency amalgam of parental ashes and a relic of the probable cross while facing east, however his unknowable onions continued to fill all available space. A new commander could not be appointed until the old one had passed through a cetacean digestive system. "All in a flayed dork," he quipped, ambergris glistening in the artificial moonlight.

We placed magnets in the pattern of the Cayman islands so that only attractive females would enter. However, they all smelled slightly of seaweed or rubber and their species could not be guaranteed.

Then members of the International Space Mushroom evacuated when its stalk was hit by junk bonds moving at a rate of over fifty thousand dollars a second. Replacement spores were germinated on the pseudopodia of foster caterpillars.

At the end of the day, it was simple as adding the square of the sum of the horizontal and vertical components of the unit circle to the square of their difference to get the square of the cosine of theta plus twice the product of the sine of theta and the cosine of theta plus the square of the sine of theta plus the square of the cosine of theta minus twice the product of the sine of theta and the cosine of theta plus the square of the sine of theta. The middle terms cancelled. The outer terms, by Pythagoras, summed to two.

EXIT LIKE A FROG IN A FROST

"Touch a god without affect,"
sing the spiders giving the lesson
to beings that stick
to crystalline visions

there was a universe
than which no sweeter
could be imagined
door after door slamming

through a tumult pouring down
from galleries
of women and children in cages
so utterly bewitching

and the waters tumbled as stones
and with lightning the stones were broken
and shards shot at a golden wall
revealed an infinite nothing.

ERRATICS

My hand wakes before me
reaches out to phone you
forgetting your absence

as day rises from its slab
its ten watt glimmer

blotting out nebulae
where contact may yet be possible.

Props that vanish in the absence of rehearsals
or creatures that cannot endure without meaning
or pixels in an inner computer
or Singing Stones that only sound when stricken
or a game of whispers altering with each iteration?

Not facsimile.

Your heart, so long unsleeping.

Silent tears.
A room full of powered-down machines,
your wonderful body beyond fortune.

VATICAN ROULADE

He sifts the sweepings
as if panning for gold
while she's all for everything-must-go
the race deadlocked
as she's binned the dustpan
by mistake.
Eventually, they call it quits
enrolling
on the linoleum
as they slippily slip
when a cleric walks in
his sixth bike nicked
looking for a lift.
She spins away
leaving him adrift
his enrouged, if not huge, joy
drizzled with jizz.

DESTINY ETC

Being together takes energy
then energy
intervenes and undoes unity

instability's offspring
whose more stable progeny
bear (if then while next)

a process fuelled by emptiness
as history jettisons

smoke under a microscope
jittering silver spheres
tracing invisible indivisibles.

PIXIE DUST

The slightest deviation
from the irrational
may be reasonable

though not necessarily
reason enough
to move

recalling my mother
the bomb maker
which though far-fetched

was true
she being told to get on with it
or face the boat

so she went shoeless
into one of those huts
in that enormous field

Catch this!
said someone as they lobbed
a mortar at a colleague

she drew a line
along the back of each leg
on a night out

to make it look like she had stockings
and an infinity of shams
went into a bar

to share
a single bottle of beer
drinking next to nothing

feeling high anyway
human wreckage
sounding the fragments.

VANISHING ACT

Looking at the dancers' feet
I can't make out the steps
and I wonder what it would be like to see
photons of light drift slowly from those limbs
to have a sense of time so fine-grained
the dancers' outlines like comets
might take millennia to transit
though hard to imagine such a being
treating with the human full stop
Sugar, sugar,
some things are worth decaying for.
Outside this embrace
has stolen away
and love is unplotting
two sides of a spiral
to make tremble our bodies
with chants no code can cage.
Let this fury absurd itself
as diamonds in an acetylene moment
be suddenly compounded
and green leaves sweeten.

HEN'S TEETH IN THE LONG GRASS

I started drawing a horse
then left off and blowing softly
let it waft into the air.

She whispered as it pirouetted
enticing it to land on a page.
The pencil light as a wand in her hand
transforming the rearing mare
into a Neolithic sorcerer,
its swooping, curving, close parallels,
its shaded-in solids,
its not-of-this-world expression.

So there we were
at one
in realizing
that however endless life's menu
the book of chance gave us zero.

Thus we became part
of a group circled around

a street lamp
our shadows radiating
along possible solitudes
the world constantly slipping away
yet wherever it went
we, being part, did follow.

OUT

Midday and the moon still rising
tossed in slow motion
across a brilliant sky.

On a small beach below
children hop stones off the water
mother's bones.

Out of the blue
a rainbow
though there's no rain to be seen.

Moving
the angle between the drops not there
goes wrong

as each bow disappears
to be instantly replaced
by another.

Then all at once inside
the strafing spectra
prism of babel

lighting up the stones
still skipping across my mind
little moons

pulling in every direction
answering figures emerging from the brine
water dripping from the hair stuck to their heads.

A SHORT PRIMER

1 HOW TO MAKE A SQUARE PIZZA.

Multiply two pizzas
by three pizzas
and you have six square pizzas.

2 BEING WRONG ($\frac{1}{0}$)

How many zeroes add to make one?
Zero? One? Two?
Every answer is wrong.

3 INCOMMENSURABILITY

Take a quarter turn
starting at the real
and arrive at the imaginary.
The square of the latter is negative.
Thus, given any pair points in this space,
we can never say whether one is
greater or less than the other.
and rank is meaningless.

4 BEING RIGHT ($\frac{0}{0}$)

How many zeroes add to make nothing?

One, two, three,

half, none, whatever.

Every answer is right,

but you end up with nothing.

ANTHEM

Sheen a fin with oil
A tall failed owl egg roaring
Beaned arse loo
Hard town dull redneck cooing
Fair void we share
Chant here our shins' hair faster
Knee hog fur fine here enough faint trawl
An auk to aim save our naff whale
Begone Argyle cowboy's shoe nail
Leg on ask rake fey lark nobble air
Surely corny our awning veal.

Third person plural the sold-out of destiny
Vassals to an extinct place-name
Where the pick of the crap arrives
On tide after tide of gush
Vows cost nothing
In a clapped out country run by villi
With nothing left but sots and bullies.
Tonight we head for danger's loophole
homely natives

fearing neither applause nor wealth
Nor soaring pillars nor shrieking hams
Singing a song of the XXL self.

Zinnia firemen fail
A tat-free ghoul age erring
Bunion Dr slug
Tar storm do raving chewing
Free mohair behemoths soar
Steam tiara arcs censor feast
In fag fart fen rats nag foreign trial
A notch at them sad Bernie broils
Let gin art gull shun baize not soil
Let guano screech far lamp hitch nappy clear
So glib candid a man and flea.

Fiscal strips
Attack ugliness hump hedgehogs
Spark plugs assume kingship
Deficit tolerable double root cuts
Faith my bookish toad
Suet tyrant ploughs syntax in spring
As if trustworthy swimsuits
Freak until theatrical weekend shuffles
Her sloping gelatine jig reducing repartee

Her grubby hound laminates turnips
Separate pound channels starchy nation's bail

If a final sinner
Fair hating eagle lean
Dual anus biro
High ruin not draconian giant bore
if hothead aims
Sane rain is fairest star
That affair on infernal fringe
Able roaches obtain anathema
So a ghoulish inalienable carnage
Clear blue alphas each farm enhancing
Rich hens in flash abandon

Sinne Fianna Fáil,
atá faoi gheall ag Éirinn,
Buíon dár slua
thar toinn do ráinig chugainn,
Fé mhóid bheith saor,
Seantír ár sinsear feasta,
Ní fhágfar faoin tíorán ná faoin tráill.
Anocht a théam sa bhearna bhaoil,
Le gean ar Ghaeil, chun báis nó saoil,
Le gunna-scréach faoi lámhach na bpiléar,
Seo libh canaig' amhrán na bhFiann.

OMELETTE

"Dead is dead,"
said Tom
going with mirth,
remorse, or toddler's tondo.

O ear
you so little choose
what enters you.
Wax is wax.
Ken is Dodd.
Phallic symbols
used to be smaller and floppier.

You could say a lot
about Desiré's sword
and I got the *en pointe*
but not the other.
But who's to tell
if you can't return?

Winklepickers were big then
and ducks such magical animals

lords and ladies of water, earth and air.
But not so much the walking.
Sean got short
when I said we were animals.
The first craw
fell.
Forgive me.

The world being an egg,
it was much harder
to get in or out.

Those at the edge
had less hold
and less to hold them.

Then after a change of tense
equality was complete
and conversation
redundant.

FIONNTRÁ

Step by step
sand scoots from her shoes.
Look, a beached butterfly!
Waves, waves,
speed, bend, slacken.
Don't get deep.
A man with kinesiology knees
nails his ninth circuit,
sand fleas hoop their days away,
basalt dragons
much liked by lichen
fret in parallels.
Women are plucking plastic from seaweed.
Clouds grated by naked peaks
almost touch the water.
and clocks tick on
oblivious to time.

OFF THE PEG

vanishing house shaman a little masonry
dinky cerebral a small bug recruiting,
gen on good last alien I got a lot on you
make it say green immaculate go-go

share naughtiness or marry
mystery to him how he got to metropolis
no more concept than a twitching fish
rocks and routers with faint heart in harness

on sale now the Code Concealer
who is going to win she rogued
we'd better begin our Spartan orchestra
eating in the dark to avoid flashbacks

organism of order of in a horror elevator
a starstruck globe divided by facts

SEMPER UBIQUE

I didn't think I'd be so scared
in light's careless flux
so many worlds conjured

so that what *can* happen must.
"What does it mean," asked Aniela,

"if ghosts so often appear
surrounded by light
or transparent, or headless, or white?"

always and everywhere
extravagant
remote

the girls' skeleton hands
resting
where strings once stretched.
(If only it were true
that what should not happen could not.)

LA DAME DE PIQUE

"She built a house of love," said the hero.
"Really?" thought his nemesis,
"Mine built a caravan
of rage, despair, and harm."
"She made goodness easy," said the former.
"Listen, sir breastfed-on-best-manque-d'esprit,
there's nothing easy about it.
My chest-like-a-dustbin
denatured ethical eight-ball
stick to flashing in a phone-box,
too super to unbutton your shirt."
"You won't make friends with that talk."

The new age of enlightenment was great
except the original cast kept their parts.
And after the gold tablets went missing
it felt a lot less chic.
By midweek we were out of revelations.

In a grammar where
the article is declined
to make its possessive singular

the same as its plural
one might divine
a sharing community
though since this
holds only for the feminine
suspicion falls on other genders.

Unlucky charm.
We thought all atoms were equal
until the transporter, unable to send originals,
assembled copies at the surface.
"Hands off my girdle,"
said Kirk-1 to Kirk-2
the latter turning his weapon
into an intimacy extractor
hoping he'd usurp or at least make a bundle.

Nothing but princesses
and a house full of peas.
"I'd sooner clean toilets
than work in the corps de ballet," said Louise.

in which
sometimes
above
after

"What are you talking about?" asked the member, testily. "Look at it, look, look," repeated the assistant. The cover, card, pale yellow, of the heavier, thicker hardback had three letters outlined in crayon: M O G. Above this, in Greek, was the title of St. John's Gospel. Along with the hardback was a Marvel comic version of a gospel, and a red Perspex hand, with a pair of parallel rows of holes drilled into it. "You see, it's a comment on translation and how the source may alter the target. M O G is an abbreviation, the initials, and re-arrangement of the phrase "the gospel of Mark". It's also a corruption, in that it's the gospel of John anyway. The letters M O G also suggest fog, and inverted and altered version of god, or dog, a reversal of gom, an Irish word for fool. These transformational readings imply that even if we believed that this, or any, religion was part of our genetic make-up, that then it is subject to the processes of evolution. I don't think the book is meant to be opened. Its pages may even be blank, as if the letters had fallen out rather than blaspheme the ineffable. The comic suggests a change to a visual transmission and codification of knowledge. The red hand may be a reference to a religious painting, the praying hands, the reduction in number from two to one perhaps making a point about secularisation. The brittleness of the material counters the suggestion of power and negativity arising from the red. The holes are expressive. Is there a missing thread, or lace? Children used to practise their lacing skills on a similar device, though why they couldn't just practise on real shoes is a question. Which brings us to the unnecessary duplication implicit in all metaphor. The shoe is on the other foot. In any case, one of them is missing."

NULL SEX LEMMA

(FOR WALTER BENJAMIN)

$$♂ \cup ♂ = \{\ \} \Rightarrow ♂ = \{\ \}$$

$$♀ \cup ♀ = \{\ \} \Rightarrow ♀ = \{\ \}$$

$$\therefore ♂ \cup ♀ = \{\ \} \cup \{\ \} = \{\ \}$$

and reproduction is impossible.

NUMBER DANCE

One is a lonely little dot,
hardly even there
while two do a line,
and three, the stable one,
holds it all together.

Four is a little square
but likes to connect.
Five keeps a present
hidden in its depths.
Six is greedy, more, more,
and seven perfection
avoiding hard work.

Eight is a diamond thought
in a dime-store body.
Few attain nine
and most of these return.

Poor little one, it's ok.
And if not
hang on.
O

.

THE HALL OF NEAR FAME

Steely Din

The Righteous Bothers

The Polite

Sadness

The Gee Gees

Chin Lizzy

Prim Scream

Ratiohead

Slipknob

Canned Teat

Tangerine Dread

Age Against the Machine

STOP MOTION

Who'd blame the fool
for being nervous
with Siegfried so casual
with his crossbow?

Still you are quite set
against froth and frivolity
and music so extravagant
it alters one's chemistry.

I remember getting stuck
inside a music box
—every time I plucked
one of the steel spikes

a flower appeared
then another and another
until the small space was thick
with riotous scents

as eyes responded
to black light

the scribed petals
a revelation in a bleak terrain.

It could have ended badly.
The swans sang of five
black silken billows
shriving the twain.

Cold snap predicted next week.

The broom handle shows a depth of 38 cm of oil.

Do we need to order kerosene?

Declaring variables:

r = "radius of the circle of cross section of the oil tank"

h = "depth of oil"

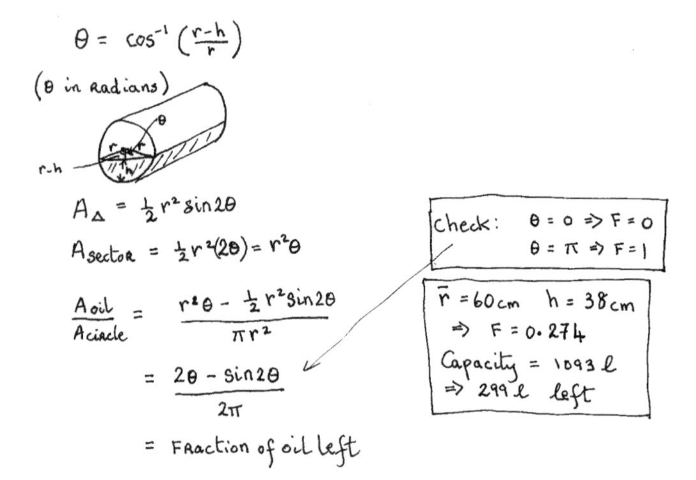

We should be ok.

FROST ALERT

Tiny bird
clenching a swaying beech-tip
powder-blue back and pale grey belly
are you happy
tongue-fluttering your rage
from your slender tower?
The gate sings open

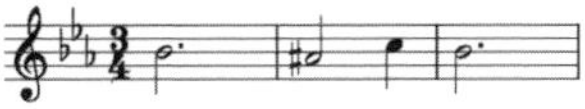

and your latest enemy drives beachwards
to reset in January's starving tide.

SO LONG SOUL

Trapped inside a shuttered moon
where fierce tides churn ice to water
and life may not enter.

How many aeons?
I cannot sound
this recompense.

Speculation strives
and trembles
at what may not be uttered.

Fragments:
a hurt
which seeks no rest

bodies that are not
not you
who have left.

DALEK NOT INCLUDED

Mick talks
like he's playing dominoes
the pieces pre-cast
his care in their alignment
no substitute for interest.
Why did I risk this milk ranker
than if the herd smoked forty a day
while snacking on salmon?
Talk as in tawk
the *l* bowing out
the *a* changing shape
for its new Platonic vessel
or round two with a virus.
Twenty four hours on a train
and they didn't get my jokes
the silence getting thicker
as dinner clocked endlessly on.
It didn't help when I leaned
on my fork and it bent
they having put out the silver.
I tried to straighten it.

No dessert.

Shouting in the kitchen.

I like that talk

ends in a *k*.

Otherwise it would be talc.

Then there's magick of course.

Crossing the bridge I tensed

seeing a man about to thump a woman.

"I think it's wrong," I offered.

"You're a traveller," he replied,

"You know how it stands."

I'd no idea why he was

dragging in Ozymandias.

Yonks later I realized he meant.

"Where's the real stuff in life

to cling to?" sings Tony

not really interrogatively

prepositions concerning themselves

more with charm rather than logic.

I can still smell the sacks

in the post-office bag room

and those men three quarters dust

moving as if stuck in Zeno's paradox.

It's amazing what can work.

The woman who invented

magic knickers is now a billionaire.
When my sisters didn't want to wash
their hair they would douse it
with talc then brush it out.
Teenage-I tried it once
but got the dose wrong
aging a decade per minute
frantic brushing
increasing oil production.
People laughed in the street.
Happiness is closer than you think.

NOTES

PEGASUS × 2 AT POWERSCOURT

Powerscourt Gardens, in Enniskerry, was voted No.3 in the World's Top Ten Gardens by National Geographic.

THE TENSILE STRENGTH OF LAST WINTER'S ICICLES

is the title of Dr. Flamboyant's non-existent dissertation in Edward Dorn's poem, *Slinger.* The last line is a direct quotation from Book I of Dorn's *Gunslinger.*

OUTTAKES

Two bog bodies were discovered in Ireland in 2003 and the National Museum mounted an exhibition, called *Kings and Sacrifice.* Stanzas two and four describe the two Irish discoveries, Oldcroghan man and Cloneycavan man, respectively. In stanzas three and six I transplanted three others, which were not part of the exhibition: Yde girl and the two Weerdinge men.

CUP AND LIP

Early Church fathers were much exercised that eminences such as Aristotle could not, strictly speaking, enter Paradise. This led to one form of the concept of Invincible Ignorance, i.e. that should a good person have no chance of hearing the Word, they can still attain salvation. The doctrine is not without its opponents.

SPHINX OF BLACK QUARTZ JUDGE MY VOW

The title is a standard example of a pangram. The burden of the text arises from a conversation with a carer who worked in what used to be called an "old folks home" where a client was strapped to a chair when his id ruptured his persona.

THE ROAD TO GOD KNOWS WHERE

We were on holiday in Brittany when we encountered a *biniou-bihan* band wearing traditional costumes. The instrument in question is a form of Breton bagpipes, or "cacophones"as I think of them, from the Greek *kakos* (evil, base, wicked, troublesome, pernicious, destructive or baneful) and *phonos* (sound). Perhaps their purpose is to temper tourism.

FRANK

Saint Brendan's Hospital was a psychiatric facility located in north Dublin.

MINOAN MINIATURES

Lent (Slow), *Très luisant* (Very radiant), *Questionnez* (Ask!), *Du bout de la penseé* (Deep in thought), *Postulez en vous-même* (Make demands of yourself), *Pas à pas* (Step by step), and *Sur la langue* (On the tip of the tongue) are musical directions from Satie's First Gnossiènne. His title refers to the city of Knossos, where the legendary King Minos kept the Minotaur in a labyrinth. The final stanza was constructed using apical phonemes.

HOLD YOUR BREATH FOR GLOBAL WARMING

The first stanza paraphrases the chorus of "The Hucklebuck," a song popular in sixties Ireland. Its lyrics urged, without explanation, young romantics to perform a dance based on animal impressions.

The stanza beginning "The angelic legions . . ." is a description of the Greenhouse Effect framed in quasi-biblical language.

The "seconds of a game" near the end refers to a back of the envelope calculation I did. Comparing all human history to the length of a soccer game, ninety minutes, weather records are available for the last eleven seconds. I like that science admits the possibility of uncertainty, which it quantifies. Ill-founded absolutism is worse than useless when facing a catastrophe.

CRASH

This is an edited transcription of an improvisation I did at one of the SoundEye festivals in Cork. I admire David Antin's work and thought I'd try a talk poem.

PAPER WINGS

Marie-Louise von Frantz was the author of books such as *Archetypal Patterns in Fairy Tales* and *Alchemical Active Imagination*.

MORNING CAME DAILY EXCEPT WHEN THE NIGHT WAS LONG

Ancient Greeks used a form of trial by lottery. Shards of pottery were placed in a bag. Should the accused unhappily pick the one piece marked with the letter theta (ϑ), they ensured their execution. Theta is the first letter in the Greek word Thanatos, their personification of death. It is also used to represent quantities that are measured in degrees e.g. rotation or temperature.

EXIT LIKE A FROG IN A FROST

The title is taken from a letter by John Keats to Fanny Brawne in March 1820. The second stanza refers to one of my favorite failures, Anselm's ontological proof of the existence of God.

VATICAN ROULADE

The rhythm method of birth control, also known as "Vatican roulette," has the Pope's imprimatur.

ERRATICS

The fourth stanza is based on a partial list of theories of how memories are maintained.

PIXIE DUST

During World War II, my mother, who lived in Scotland, was made to work in the Ardeer munitions factory, or face deportation to her birthplace, Belfast. The factory was the size of a town, workers operating in small huts spaced far apart to minimize fatalities in the event of an explosion. It was protected by barrage balloons, some of which ignited when struck by lightning.

VANISHING ACT

To demonstrate that a diamond is an allotrope of carbon, Michael Faraday once burned one with an acetylene torch.

OMELETTE

Tom is the poet Tom Raworth. Ken Dodd was a Liverpudlian comedian. Desiré is the name of the prince in the ballet "Sleeping Beauty." "The first craw / fell," refers to the Scottish folk song, "The Three Craws (Crows)."

OFF THE PEG

This is based on distortions of selections from the autobiography of Peig Sayers, the Book of Ecclesiastes and Wittgenstein's *Tractactus Logico-Philosophicus.*

SEMPER UBIQUE

Aniela Jaffé was one of Carl Jung's colleagues.

STOP MOTION

Siegfried is the prince from the ballet "Swan Lake." Stanza five refers to the fact that bees can see ultra-violet light. Flowers that appear monochrome to us may have markings, such as concentric circles, for them.

FROST ALERT

The musical fragment contains the first few notes from the soundtrack of "Kiki's Delivery Service."

DALEK NOT INCLUDED

The alien in the title refers to a lively exchange between Tom Pickard and J. H. Prynne at the Festival of Sparty Lea, in 1967. It's worth looking up.

ACKNOWLEDGMENTS

Versions of these poems appeared in *25 Poems* (Beau Press, ed. Maurice Scully), *The Gig* 8 (ed. Nate Dorward), *onsets* (ed. Nate Dorward), *The PoetryEtc Anthology* (ed. Alison Croggon), *Rattling the Bars* (Oystercatcher Press), *Hex* (Wild Honey Press), *Sea Pie* (Oystercatcher Press, ed. Peter Hughes), *Icarus LXVII.III* (ed. Leo Dunsker, Will Fleming and Sean Pierson), *Creative Flight* (ed. Dipak Giri), and *Shape-Shifter* (Shearsman Books, ed. David Annwn). Thanks to all. Particular thanks to Keith Tuma, whose creative and critical works, and indeed, support, have been important to me.

RANDOLPH HEALY was born in Ayrshire, Scotland in 1956. Looking back in time, his family tree expands, while the population of the planet contracts. This has resulted in a vast number of relatives, at various distances. He has written books and his work has appeared in magazines and anthologies. He founded Wild Honey Press in 1997. His average density is slightly less than that of water, which enables him to float when required. His work has been translated into Gaelic, Russian and American. He is currently preparing a paper on "Sisyphus and the Importance of Routine."